Book Ends for the Reader

Topic: Me and My Family

Notes to Parents and Teachers:

The books your child reads at this level will have more of a storyline with details to discuss. Have children practice reading more fluently at this level. Take turns reading pages with your child so you can model what fluent reading sounds like.

REMEMBER: PRAISE IS A GREAT MOTIVATOR!

Here are some praise points for beginning readers:

- I love how you read that sentence so it sounded just like you were talking.
- Great job reading that sentence like a question!
- WOW! You read that page with such good expression!

Let's Make Paired Reading Connections:

- First, read the fiction text, ***A Trip to Grandma's House*** by John Wallace.
- Next, read the nonfiction text, ***Do Animals Have Families?***
- Discuss how the pictures in the books look different. *One has drawings. One has photographs.*
- What are both books about? What a book is about is called a topic.
- After reading both books, what do humans and animals have in common?

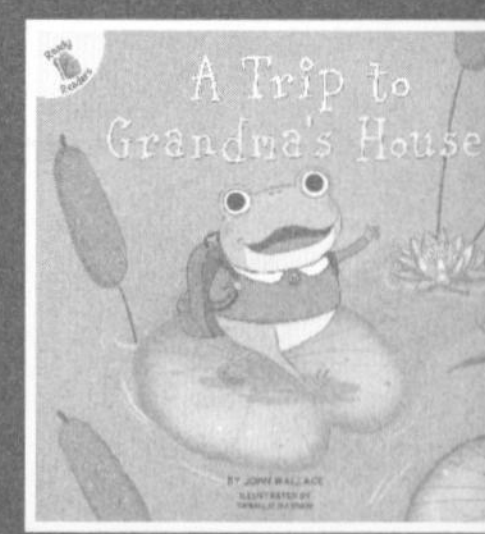

A Trip to Grandma's House

Table of Contents

rourkeeducationalmedia.com

Can you find these words?

groom

guard

pod

pride

Do Animals Have Families?

Some animals have families.

Elephants live in a family.

The mom is the boss.

Lions live in a family.

The dad protects his **pride**.

Chimpanzees live in a family.

They **groom** one another.

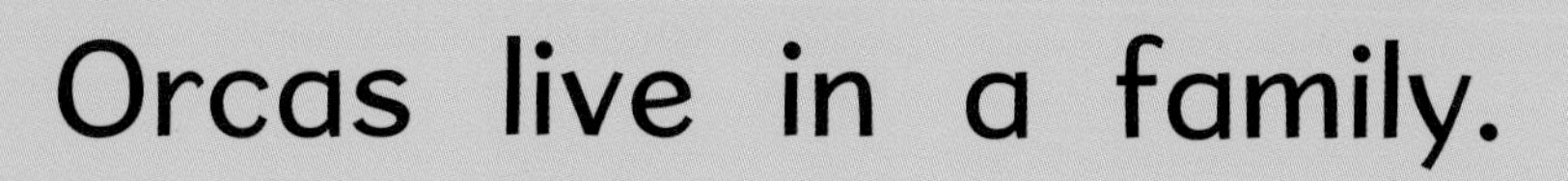
Orcas live in a family.

Their family is called a **pod**.

Meerkats live in a family.

They take turns standing **guard**.

Did you find these words?

They **groom** one another.

They take turns standing **guard**.

Their family is called a **pod**.

The dad protects his **pride**.

Photo Glossary

groom (groom): To brush and clean an animal.

guard (gahrd): To protect someone from harm.

pod (pahd): A group of a certain kind of sea animal.

pride (pride): A group of lions.

Index

About the Author

Michelle Garcia Andersen lives with her husband and three teenage kids. Her dogs are also a big part of her family. She lives in southern Oregon in a home full of laughter.

www.rourkeeducationalmedia.com

PHOTO CREDITS: Cover: ©tratong; p.2,8-9,14,15: ©ThomasDeco; p.2,12-13,14,15: ©nattanan726; p.2,10-11,14,15: ©Tory Kallman; p.2,6-7,14,15: ©Teresa Moore; p.3: ©prasit chansarekorn; p.4-5: ©mlal33

Edited by: Keli Sipperley
Cover and Interior design by: Rhea Magaro-Wallace

Library of Congress PCN Data
Do Animals Have Families? / Michelle Garcia Andersen
(Time to Discover)
ISBN (hard cover)(alk. paper) 978-1-64156-205-8
ISBN (soft cover) 978-1-64156-261-4
ISBN (e-Book) 978-1-64156-309-3
Library of Congress Control Number: 2017957903

Printed in the United States of America, North Mankato, Minnesota